"I'll grab a fiddle," I yell.
"You grab a friend.
We'll have a good time!"

People get in lines.

I pluck the fiddle strings.

“Shuffle to the middle,” I sing.

“Clap your hands and
shuffle back.”
People do this simple step.
No one slips, stumbles,
or tumbles.

They swish, spin, and dip.
They dip to their friends.
They spin and stomp as
I fiddle.

Do people want a fast tune?

You bet!

Some people miss a step.

They tumble into a pile.

I stop in a bit.

People get apples as I sit.

I sample some cake.

Soon I grab a fiddle.

Fiddle, faddle, diddle, fiddle!

I sample a tune.

People run to get in lines.
They run back to the middle.
They are set to have fun.

The End